If You're Considering Mobile – This is a Must Read!

Mobile Search....

Special Offer Inside

....the Rush for Mobile Gold

by David J. Dunworth
Author of *The Entrepreneur's Bill of Rights*, *C.O.M.P.A.S.S. Points*, *The Newbies Guide to Email Marketing* and more...

Entrepreneurs, Agencies & Small Businesses Can Greatly Benefit from Mobile Today!

Mobile

Search....

....the Rush

for Mobile Gold

by David J Dunworth

Author of The Entrepreneur's Bill of Rights, C.O.M.P.A.S.S. Points,

Look for the Special Offer!

Copyright© 2013-3-2016 Marketing Partners LLC

ISBN – 13: 978-1493617654

All rights reserved.

No part of this publication may be reproduced in any manner whatsoever without expressed written permission from the author.

This book is designed to provide accurate and authoritative information regarding the subject matter covered herein. It is sold with the understanding that the publisher and author are not engaged in rendering legal, accounting, financial or other professional advice and/or services. If legal advice or other expert assistance is required, the services of a competent professional person should be sought.

From a Declaration of Principles jointly adopted by a Committee of the American Bar Association and a Committee of Publishers.

The author and publisher wish to make it known that the information expressed herein is the opinion of the author and not necessarily the "final word" on any given subject. However, it is one person's view and biographical perspective. Because of possible unanticipated changes in governing statutes and case law relating to the application of any information contained in this publication, the author and publisher disclaim all responsibility for the legal effects or consequences of any action taken in reliance upon information contained in this publication and cannot be liable for any loss as a result of the application, directly or indirectly, of any information provided in this publication.

Although every precaution has been taken in the preparation of this publication, the publisher and author assume no responsibility for errors or omissions.

DEDICATED TO

The Overseer of Order, who makes me want to be a better writer

Table of Contents

Preface .... - 11 -
The New Gold Rush .... - 13 -
Therein Lies the Challenge .... - 28 -
Think Local .... - 37 -
Making the Case .... - 42 -
Big Disadvantages .... - 52 -
Mobile is Better than Wide-Screen .... - 61 -
Mobile Makes Sense .... - 62 -
Quality not Quantity .... - 66 -
Business Mobile Sites Popularity Up .... - 72 -
Mobile Apps Down – .... - 72 -
Numbers Don't Lie .... - 82 -
Mobile Marketing Statistics 2013 .... - 83 -
The Marketing of Mobile and the Opportunity .... - 100 -
Preparedness Meets Opportunity .... - 108 -
Unparalleled Opportunity? .... - 117 -
Read Just a bit further! .... - 117 -
About the Author .... -129 -

ACKNOWLEDGMENTS

Dr. P. Z. Pilzer & Jay Abraham, whom I consider my two mentors although we have never met, and whom I have borrowed so many great ideas

Ian, for always putting up with my whining

Samantha for her terrific cover design

Alex to for putting up with my ever-changing priorities

My great friend MAK (Montgomery Alexander Kaine) for his unwavering loyalty and encouragement

Preface

Look around and what do you see? People are making their way to work, to the store, they are off to the dentist, the audiologist, the coffee shop, the spa or maybe even the fortuneteller. They are all doing so while looking at their phones. Their eyes are glued to the screen, fingers clicking the keys. Their mobile device has become another appendage.

The cell phone, or smartphone as we call it now, is with us constantly. People have been captured on video strolling in a shopping mall and tripping into the fountain because they were staring at their smartphone. Busy people walk into traffic while being immersed in something on their phones. They bump into poles, trip on their own feet and make silly errors in judgment as a result of them.

Smartphones also grow relationships, business, even save lives. Our mobile society cannot seem to get enough of the digital instant gratification smartphones, and other mobile devices provide. It has been documented that teens would rather pass up receiving a car instead of having to give up their phone. People have gone on record stating they would give up television, new clothes and even giving up chocolate instead of the loss of their smartphone.

Nearly everyone with a smartphone would probably give up something just to keep their devices. There are more mobile devices in use than there are people on the planet. That is a lot of mobile devices. Did you know that of all the population in the world, more people own a smartphone than own a toothbrush? All and more of these statements are true and have manifested because the world has a love affair with their mobile devices.

If you have yet to do so, these are reasons to get involved with mobile. The new gold rush is already happening; it is time you got involved.

The New Gold Rush

The world's eyes are on mobile. Mobile technology is at its greatest stage of growth, and the tsunami of mobile is growing larger still on its momentum. While the focus has been on mobile for some time, An astounding 11.8% of all websites were still not mobile compatible by the end of 2015. Although 2016 has seen a push in that direction, not all "experts" believe responsive design is the cure-all people say it is.

Why get mobile? These outdated, static sites cannot be viewed on a mobile device without scrolling left/right, pinching and scrolling up/down. Mobile websites change all that. With a global society focused on mobile devices and performing a search on their smartphones and tablets daily, it makes little sense why businesses worldwide have yet to figure out mobile's real potential. It is amazing they have yet to go mobile. You now have that ability.

Mobile phones are no longer considered new; their full capabilities have yet to be mastered, but that is quickly changing. As technology is whizzing faster and faster, it is bringing with it new and improved devices to market almost daily. We have mobile watches, and even they are not considered the latest thing. Google Glass, the mobile Internet eyeglasses had been selling in beta, but fizzled. There's a new player on the scene, In fact, there are more than 50 different wearable technologies out there at present. Some will make it, some will not.

Mobile makes sense. Mobile search is the next wave, and it is hitting stores around the world in the tsunami-fashion of which I spoke earlier. Statistics show (you will see lots of them further into this book) that mobile search eclipsed desktop search before the end of 2015, and that trend will only accelerate over time.

Ask yourself this: what will a visitor see when they search your site on their mobile device? The answer you get can make or break your business. Consumers quickly move to a compatible site (of your competitors) if yours is not viewable without difficulty. Today's mobile user has become so instant gratification starved, the wait time before they move away from a site is less than 8 seconds. Not much of an attention span.

It does not matter whether you are a Small Business, an

Entrepreneur, a Marketing Agency, Business Owner/Manager, Local Marketer or whether you are an Emerging Enterprise, your customers deserve a good mobile experience. You are worthy of a mobile site that does what it is supposed to.

There isn't a marketing agency in the world that should not be offering their clientele mobile solutions and not just a responsive website. Every business needs to be visible on mobile devices because the world is already checking out their sites on a mobile device or an app.

Will the person searching for your business on a mobile device cause them to spend time on your site or make them rush to some competitor site? Check yours again and then come up with the answer for your own business. Mobile makes sense.

A NOTE of Particular Interest?

If you are an Entrepreneur in search of the right opportunity, you have come to the right place. There is a special offer further in this book that you will not want to overlook.

DAVID J DUNWORTH   MOBILE SEARCH.....THE RUSH FOR MOBILE GOLD

The Technology Gap

We have seen it all before. The world has witnessed many times how technologic advancement has replaced and even eliminated entire industries. It did not matter if the particular product or service was the absolute best in the world. Once technology had advanced and narrowed the gap between existing and new, it often left inferior tech to wither and die. Either the accessibility, ease of use, cost of distribution, the "making life easier" factor or any other way you want to slice it, the advance of said technology wiped out or crippled industries on a global scale.

The pace of shifting priorities from an old product to a more advanced model often creates a tidal wave of economic activity and growth. It also devastates outdated industries. Once a new technology is recognized, it is only a matter of time before it completely engulfs what it is replacing.

The ***technology gap*** is the time between current technology becomes redundant once newer technology reaches critical mass and overtakes the inferior product.

The technology gap happens in things other than products, but for today, let's stick with see, touch, feel items that we have experienced or soon-to-experience the technology gap being closed.

Some goods and / or services take months or years to close their particular gap while others close when we are not even aware they are dwarfing the product or service which it is replacing. This is known as *filling* the technology gap. Not sure of what I speak? Here are a few examples to get you thinking about technology gaps and how it occurs.

- ❖ **Ever Heard of the Buggy Whip?** Of course, you were not around in the days they were used, but I know you have heard of them. Before the end of the 19th century, local transportation relied almost entirely on horses. The horse and buggy were widely used to get around. Nearly every family had at least one of each, and carriages were available for hire in every community. No buggy was without a whip; it came with the package and was a vital component in the movement and motivation of the animal. The driver simply had to nudge the horse or mule with the whip, and a well-trained animal would go forward, left, right and even backward. A stubborn animal needed more "direct" motivation, thus the whip's intended use.

    Then along came the "horseless carriage." Within 30 years, every buggy whip manufacturer worldwide was completely out of the buggy whip business. Those remaining in view are in museums.

Those with enough foresight and intestinal fortitude switched production to riding crops, saddle gear and other equestrian goods to survive. Unfortunately for the overwhelming majority of buggy whip manufacturers, they disappeared from the landscape forever.

- ❖ **Remember the Kodak Company?** Are you aware that Kodak invented the digital camera? The truth is stranger than fiction. In the mid-1970's, a man named Sasson invented the digital camera for Kodak. This engineer brought an idea to life. However, the focus of the company was what it started out to be way back in 1900; film. In fact, the Brownie camera was made by Kodak and sold for about $1, with another 15 cents a roll for the film.

In the year 1999, Kodak produced roughly 70% of the world's supply of film. It had annual sales of more than $1 Billion, with a B.

In 2011, Kodak filed for Chapter 11 bankruptcy. They failed to recognize Moore's Law (the technology gap being filled). Today they are nothing more than a memory.

- **The Carburetor** – The combustible engine has run on a mixture of gasoline and air since its invention and was the component responsible for that combining of "gasses" to make the motor run. Two Barrel, Four Barrel, Ram-Air and Single Barrel carburetors were on each and every engine coming out of Detroit and worldwide manufacturers as well. Long before Henry Ford streamlined the auto building industry with his production line techniques, the carburetor ruled the auto industry when it came to making an engine run.

America's cars burned gas like there was no tomorrow, because, in a manner of speaking, gas was so cheap that morning seemed a long, long way off. Gasoline in the US cost only 29 cents a gallon in the late 60's, so the size and weight of cars back then didn't matter; gas was cheap. Even though Detroit was building these land-yachts faster and faster each year, it did not seem to matter. America loved to drive in style and comfort. It was affordable too.

The oil embargo of the early and mid-1970's woke up the western world and the auto industry, as the first time the thought of running out of oil came into clear focus. Loss of oil would mean the elimination of gasoline; there had to be a way of making our reserves last more than the predicted 60 years the pros said we had in the Earth. This urgent need caused a mind shift of tremendous proportions.

Cars began to shrink, got lighter and more streamlined all in a way to increase fuel consumption, but it did not make much difference.

The Land Yacht was gone for good. Bumpers turned to plastic along with a thousand other parts. It did not help. Thankfully this tale of woe trickled all the way down to the carburetor.

[Jonas Hesselman](#) of Sweden invented the earliest version of gasoline fuel injection way back in 1925, but it was ignored by Detroit (diesel fuel injection was developed in 1902, by French Aviation Engineer [Leon Levavasseur](#), but wasn't considered viable for the automobile). These men had the idea of creating a system of calibrated fuel and air mixture that was precisely consistent and never varied and pumped into a cylinder at the precise moment of firing. The world's consumers could get from place to place without having to worry about the care and feeding of a horse and the ongoing repairs of the buggy.

There we have it; the fuel injection system. Ignored by Detroit and the world's automakers until their backs were against the wall when everyone feared the end of oil, it suddenly was perfected and put into use. Help it did. Fuel consumption doubled, and then doubled again. Partly because of improved fuel consumption via the fuel injector as well as new discoveries in oil reserves in places yet to be explored.

Maybe employing the systems of the fuel injector, smaller size and weight of the vehicles and national speed limits set at 55 mph, they would help extend our oil reserves to keep us in gasoline longer than a measly 60 years, as was forecasted in the 1970's.

Smog equipment did not even exist on cars until the late 70's. Even with all of the necessary smog equipment added to cars these past thirty-five years or so, we now have the safest cars ever made. These marvels run more than 40 miles a gallon and don't impact the environment anywhere near the way they did in the early twentieth century. Today's cars are even smaller, the bumpers are still plastic, along with those thousand parts.

Another fuel related technology gap is currently being filled to take yet another leap forward; hydrogen, all-electric and natural gas vehicles. These clean fuels will ultimately replace whole industries. Once technology brings about affordability to create demand at critical mass, greedy oil companies and even more mercenary lobbyists will be wringing their hands. That day will not come too soon.

Moore's Law is having another effect regarding the automobile; the driverless car. Test markets are already popping up across the country as companies race to be the first actually to bring them to critical mass.

Additional Examples

- ❖ Other technology gap-filled examples are showcased by looking at the **Vinyl Record Industry,** replaced by the cassette player and tapes, and then the **Compact Disk**. CD's have since been replaced by **digital download** of not just artist's albums, but individual songs.

- ❖ **The Video Cassette Recorder** by the DVD player, to the DVD recorder, **Videotape** to the DVR; the list goes on. In all of these cases, whole industries were either brought to their knees or wiped out.

- ❖ **Voice mail** replaced the voice recorders (message machines) for your home phone, then the **cordless home phone** replacing the cord in the wall phone. The mobile phone is

minimizing the home phone due to the ability to call long distance for the same price as minutes on their mobile plan.

Unlimited national calling brought about cost-saving convenience, and also predictability of cost per month. The family now has a way to budget their phone usage. It is the texting by teenagers that have parents still on the edge of their checkbooks.

- ❖ ***Typewriters*** were replaced by the ***early word processors*** and shrank the giant pools of secretaries outside of the corporate executive's offices. More output per typist meant less payroll headcount.

- ❖ ***The personal computer*** practically eliminated the need for a secretary; even companies as large as Ford Motor Company had its mid-level management perform their own secretarial work as early as the 1980's. What once took armies of clerical personnel can now be done by a handful.

- ❖ ***The personal computer*** blossomed into the laptop, shrinking sales of stand-alone boxes significantly, and making work portable. Today we have ***tablets*** which are taking the place of the portable laptop computer. Of course, the smartphone is eclipsing both.

- ❖ ***Border's Books*** was a national chain of bookstores that offered the complete range of genres, magazines, you name it. The leadership scoffed at the electronic readers and said "they will never replace a printed book. Within a few years of Amazon's Kindle™ eReaders for sale, Border's Books were shuttered everywhere.

* **Cloud storage** is eclipsing the large server companies need to harness all of the world's data and software, reporting, legal and compliance and other documentation and systems analysis needs.

* **Smartphones** have put the power of the computer in the palm of our hands. Smartphones are capable of nearly everything a desktop, laptop or another computing device can do, if not better. Moreover, it fits in our hand. The tablet is assisting the smartphone in eliminating the desktop computer further than before. No longer does one have to be tethered to a desk to get work done. Heck, the wait for reviewing documents or jumping in on a conference call is as close as the smartphone in your hand.

That's a whole lot of gaps filled by new technologies, and it will not stop anytime soon. Innovation is at its peak, as the Internet has opened the world of information like never before. Tomorrow's consumers will only know technology, and making jokes about how we could survive without the technology of the day. They will wonder hat black and white television looked like, how did VCR's work, even why cars used to have to be driven by people.

Today's consumers are getting younger every day, and a vast majority of them were born in the digital age. Baby Boomers (many of which are averse to computers, mobile tablets, and even smartphones) are also retiring in huge numbers. This age-related gap is closing, and young people are tech-oriented, speaking something called "geek-speak."

Therein Lies the Challenge

Recognizing a technology gap in the early stages and taking advantage of the shift in technologies in which people choose offers an opportunity to make significant sums of money, as these visionaries have a foothold on the market before it is widespread. That is how the logging barons, railroad and utility owners, land-barons and every other type of early visionary made vast fortunes, while others just went about their daily lives in sublime misery.

The Great Depression of last century saw millions and millions go broke, starving in the streets looking for handouts, while those with vision became millionaires and billionaires. Our Great Recession of the last decade, mixed with advancing technologies is and will continue to create more millionaires than ever before. Those who can recognize a global trend and capitalize on it. Unfortunately, if you are not the one with the idea, passion and ability to find funding, you will be on the sidelines watching.

By the time critical mass occurs (when a significant portion of the population catches on), it may be too late to profit, no matter what new technology hits the market.

Unfettered growth occurs in all industries, but mobile is the fastest growing industry in the history of the world. The field is crowded with companies and individuals that have developed their brand visibility, have penetrated the market and have created loyal customers. They have developed deep pockets, and have momentum on their side. Competition is fierce at the critical mass stage. Mobile is positioned to go beyond critical mass in the next few months if it has not already. If there is any interest in making big money in mobile, you need to act now. Some say it is too late, but I do not subscribe to that negativity.

As in most things, a significant amount of today's digital marketers seems to be looking for a panacea, a magic bullet. An easy fix to bring the profits to their door with little or no work or investment does not exist. It does not happen that way.

Marketing and advertising no longer work in the traditional sense either. Consumers are far too savvy and are tired of being bombarded with advertising and marketing hype, as has been the case before now. The failure of newspapers, radio stations, telemarketing firms and many other industries have been negatively affected by the change in buyer behaviors.

Technology gives the consumer control over what he chooses to watch, listen to or read. The consumer is now in charge because they research that they want, and pick the vendor to which they can relate. They do it online.

Big corporate marketers have been playing at this mobile marketing thing for a few years now and still don't have the magic bullet. You may hear or read that mobile advertising does not work. Ad blocking software is preventing slick marketers from making a deep penetration in the mobile ad sector. Websites are now "responsive," allowing mobile viewers to have the benefit of viewing the entire internet site. Such is not the case, as there is far too much information that goes unviewed and is very expensive to move to a mobile site. Furthermore, a complete site on mobile would take too long to load. Millennials have no tolerance for slow-loading sites on their mobile device. Then there is the misunderstanding concerning responsive and mobile optimized, They are two different things.

Secondly, mobile advertising is still in the early growth stage, destined to become the primary method of reaching customers in the next few years. The march to mobile is happening, and more and more companies are forecasting a sizable increase in mobile budgets in 2016/2017 and beyond.

> "If your plans don't include mobile, your plans are not finished."
>
> -Wendy Clark,
> Senior VP, Integrated Marketing
> Coca-Cola Companies

Marketers have yet to learn that there might not be a single fix for all of the desire amassing to "reach the mobile consumer." Location-based advertising, QR codes, banner ads, click-through alerts, SMS targeting, pay per click, mobile couponing, digital punch card loyalty, online presence; these and more all have traction. It has to begin somewhere because the numbers cannot be ignored. Mobile is that somewhere. Mobile is that unique piece of the puzzle, the magic bullet (at least for the time being) for what marketers have been yearning. Not enough of them have realized it yet for the local marketing of products and services, or don't understand that content marketing is not sufficient.

Mobile is right in the middle of yet another Technology Gap of sorts right at this very moment. Think of it as a tsunami within a tsunami. Sure, the mobile phone has been around for a while now; from its early stages of a big block in a bag to a brick in our hand, eventually getting smaller and more sophisticated at every step. The innovation of all things mobile seems like it has been going on forever.

Cameras have been added, along with a bunch of apps and cool things to do with them. Video and HD high pixel cameras make some smartphones rival in picture quality as an SLR film camera. Batteries last longer than ever, and there is always more innovation and upgrades on the horizon, no matter the direction. The constant improvement and enhancement of mobile's capabilities make it constantly in flux. That is what makes mobile so exciting.

As for first mobile marketing, the only thing that seemed to get any traction was SMS (short message service), or texting.

Text marketing strategies were early winners, and like anything that is only half-understood was quickly "overdone." Unregulated texting offers (SPAM) turned just about every consumer against mobile marketing altogether. Banner ads just didn't take off, QR codes were misunderstood, and the concept of mobile advertising and commercialization was floundering. Retail was just about the only industry making something of mobile marketing tactics. Someone figured out that digital couponing can bring in customers willing to spend. At first, retailers had consumers print out a coupon from their desktop computers. Now merely showing their smartphone with the downloaded coupon is quite enough to score the discount or another special offer.

Giant Step Forward

Mobile Search is the foundation, the Rock of Ages s to speak, that the mobile marketing industry has been failing to pay enough attention to until the past couple of years. What traditionally seemed superfluous is now the primary way for businesses to "get found." Also, companies used marketing methods for decades that no longer work. Reach the customer by bombarding them with push messages? Shout at them on television, with outlandish claims that didn't make sense. Telemarketing at dinner time was one of the worst methods used, and people grew sick and tired of interruption techniques. Traditional marketing stopped working because the consumer became research oriented and self-directed, and we learned that users would rather find the businesses online with which they cared to do business. They would rather skin themselves alive than to get another telemarketing call, but they never forgot the firms that tried to bury them in texts. They despise them to this day and will never forgive them. Companies that performed in this fashion have been sidelined by the consumer, who like an elephant, never forgets.

Marketers, who have become superb at the 'Inbound Marketing" strategies of content, relationship building and nurturing failed to pay attention to mobile too. Everyone was busy focusing on "content" and failed to realize that mobile search was taking center stage in the consumer research area of business. Fortunately, video has grown so important for consumers that it is overtaking written content. The fact that YouTube is the largest search engine sums up how customers feel about video technology.

Consumers are never far away from their mobile devices and are surfing the net regularly for information and entertainment. Incorporating mobile into every business' marketing strategy will change their world, but most have not been educated enough yet.

We are a mobile society, and so is the world beyond the U.S.A. Getting around and finding the need to "search" a business on the smartphone or tablet is what the consumer is busy doing, and the companies that understand this prime directive will win.

Mobile search is the answer. Many believe search has moved away from By being able to get found on a mobile device easily and with unbridled viewing, local businesses can grow revenues and advance their brands through mobile. Take a restaurant, salon, hardware store, tow truck company, bar, grill, store, or you name it, someone is viewing their site on a mobile device. Most of the time mobile searching is for only four things. I like to call them the Big 4. Location & directions, special offerings or information, hours of operation and a way to contact them quickly.

Mobile Search is the key to growing a customer base. If a business has a mobile website, all of the pertinent information is face-front, without any effort to hunt it down. If a potential customer performs a mobile search and your site is not viewable, they quickly move on to the next one on the long list of competitors. Complete websites take an enormity of time loading, are difficult to find the Big 4 components of which most people are in need. Mobile search finding mobile sites wins.

Google, who does not have an apparent monopoly (just kidding) in search (94.6%) though it dictates the way in which everything is found on the Internet. Yahoo and Bing are "me too" brands, but there is no doubt that Google sets the rules. Did you know that Google decided in early 2013 to begin punishing websites that did not also have a mobile device component to them? Yep, if your site cannot be easily viewed on a mobile device, your search rankings suffer.

Another clear indicator that mobile search is the new "must have" is by taking a quick look at Google + rankings. Google rewards companies that have an engaging current website and a mobile site. Social media connectivity on the go is a big part of the Penguin update that occurred several months ago, and more than 70% of social media viewing is performed on a mobile device. Shouldn't every company invest in social media also invest in mobile? I think so and so should you.

Think Local

The key to unlocking marketing is on the local level. Mobile search is at the heart of local marketing. Identifying and connecting the consumer with a business entity is what it is all about, and the local connection is most important. Not paying close attention to this segment of the market is utterly foolish, yet some companies are stuck in the traditional stratagem and don't understand where their customer works and lives; locally. They think that because whatever marketing tactics they attempted that collaborated in the past will work in the future. That is an insane notion. The future is here, and old school marketers and companies are stuck right where they started; behind times using outdated methods, and avoiding the use of modern technology. Take a look at what Albert says about repeatedly beating the dead horse.

Image Credit: Leader-Values.com

Fortune 50 companies are focusing on the local marketing effort at a ferocious pace. They are investing millions into local strategies, and they are not squeamish about investing in mobile search. This is one of the great reasons why local marketing can best serve the community by working with small business providers like you and me. Mobile search is so important that a company would be foolish not to have a mobile website.

Some mobile "optimized" desktop sites may look good on a mobile device screen, but there is excessively too much information for the viewer to absorb. Imagine ordering a mobile website to be built and having all of your company's content on it. Crazy expensive. Some webmasters or designers may claim "responsive design" (meaning the site will automatically resize for mobile) is sufficient. Almost all of the time you will still have to scroll up/down and left/right because images may not resize correctly, graphics and headlines will weird out, and other troubles with which searchers do not wish to contend.

Here's the rub. Searchers are not looking for the entirety of your business; they are seeking to connect quickly. Responsively designed desktop sites sound like a good idea until you search for a company and have to hunt for the information you are after.

Regardless of whether you are an inbound marketing agency, local business owner, independent business consultant or entrepreneur, your customers need to be mobile optimized. Every local business does business with the local community in one way or another. Make it easy to business with you, because they will buy from someone. If they search for a product or service, or merely an industry and your website is on the listings, you should want to make sure your potential customer can access the Big 4 with no trouble at all.

Making the Case

Search is no longer restricted to the desktop or laptop; mobile search eclipsed desktop near the end of 2015. It is estimated that mobile search is as much as 70% of the total. People in the U.S. have already surpassed desktop with mobile, and worldwide, mobile constitutes more than 30% of all search. People are on the move more than ever, and working away from the office is quickly becoming the norm. The rationale of mobile search is quite different from traditional search, as the information is accurate to location, offerings, hours open and quick means of contact. Waiting for a complete website takes far too long. Having to sort through a full sized site on a smartphone is just too difficult.

What makes such a remarkable difference in the way people use the Internet is the fact that we have become a "mobile" society. Did you know that there are more hand-held devices on the planet than there are people? It's true; just think about yourself: Do you have a smartphone? How about a tablet? An eReader? What about the new hybrid devices? It is an easy conclusion that things are, and will become even more and more mobile oriented, soon leaving desktops and laptops in their wake. All the more reason why every business needs a mobile website.

Image Credit: Eyeson-Mobile.com

Despite the fact that desktop website design has become "responsive" (a full sized website shrunk to the size of a smartphone) doesn't present anywhere near as user-friendly as it should. Also, do you need your entire website available for a mobile search? People search differently on mobile devices than they do computers. With a desktop or laptop computer, the screen is large, allowing the user to navigate with ease. A tablet or smartphone, on the other hand, has so little space that the conventional design just doesn't work well, no matter what the web designer says.

Don't believe me? Prove it to yourself and search for any business on a laptop or desktop, and then view it side by side on a hand-held device. If they do not match up with the size of the screen, the website is not mobile optimized. In fact, eight out of ten sites are NOT mobile optimized, and the same ratio holds true for businesses without mobile internet sites. Eight out of ten businesses today worldwide do not have a mobile site.

Businesses are at the mercy of their competitors that have embraced mobile. This fact makes the "mobile" opportunity so tremendous.

Every business in the world needs a mobile website if they are to compete in their industry on a local level. Local marketing is the bastion of opportunity and commands the greatest focus, mainly because small businesses must fighting fiercely for survival. It is no longer enough to place a poorly conceived ad in the book with the yellow-paged listings, nor is it to spend big bucks with the blue envelopes and pray for business. Although there are thousands of companies still paying for such outdated marketing in print AND online, it is clearly not the wave of the future. Once the paper directories were the primary revenue generator for big book publishing companies, but they are now holding on to sustainability with every ounce of their being.

The economic storm we are in is not going away anytime soon, and competition is growing in fierceness. When it comes to marketing, online is apparently taking over traditional marketing methodologies.

Sure display ads, billboard advertising, trade shows and newspaper and magazine ads are still financially viable; however, their revenue production is eroding mainly because of the online shift the consumer has created.

The daily newspaper is becoming obsolete, so print media no longer is viable. People do not want to be shouted at through the radio (satellite now dishes up commercial free audio entertainment). So many households have abandoned the home phone because they are tired of being bothered by telemarketers (usually at the dinner table), tired of watching television commercials (thus the popularity of TIVO and the DVR), pushy salespeople in department stores and even door to door (direct) sales.

The consumer is now in charge. Today people are searching online for the products and services they need, and for the companies with which they wish to develop a relationship. Marketing is not adapting to this shift fast enough, presenting opportunities to those who are tuned in, especially the Millennials.

Some of that type of search is still being performed on a desktop device, but that is changing daily. Widescreen search, where one can take time surfing the net at home or at work no longer holds top-of-mind search. Multiple devices are being used today, such as Smart TV and tablets or smartphones in tandem. The online movement is shifting from wide screen marketing to mobile for more immediate information, and mobile surpassed desktop search at the close of 2015. That gap is widening. The prime objective of marketing is to compete through mobile, for the simple fact that mobile search is how people are using the Internet while on the go and at home and work.

Don't misunderstand me; desktop search is still a vital option, as people search and research information about intended purchases, find in-depth information about particular topics and product or service information.

There is no limit to the items being explored, and the data online grows by millions of pages weekly. Smart marketers are designing their offerings, content and contact information with mobile in mind.

Although it has diminished, desktop/laptop search will remain extremely viable. Widescreen viewing of in-depth information is much easier to experience, and a complete site is still quite indispensable. When someone is surfing the web for information and researching a particular product or service, desktop computing makes more sense. However, people on the move and using mobile devices to search are doing so for significantly different reasons.

A company website viewed on widescreen is an expectation of providing access to all of the enterprise information. Searchers are delving deeply into sites for all sorts of information that may not be on the home page or under one of the page tabs.

Image Credit: Wikipedia

Mobile search,

on the other hand, is entirely different.

First of all: viewing a static site on a hand-held device just does not present very well. Searchers have to scroll left/right and up/down just to read the thing. Pictures that don't resize to fit the device block a majority if not all of the screen. Mobile sites work as intended, especially when Eyes On Mobile technology is utilized. The results are guaranteed.

Secondly: the mobile site must be designed for search with fingertips; there is no mouse.

**Next:** people on the go want to find out where a company is located and how to get there, their hours of operation, phone number and if they have a particular product or service. They want and are expecting the critical information to be readily available and accessible instantly.

Furthermore: Mobile sites are designed for rapid response to the needs of the busy consumer. The home or landing page of a mobile site has all the necessary information with the touch of a button.

DAVID J DUNWORTH		MOBILE SEARCH.....THE RUSH FOR MOBILE GOLD

Finally, even Google is in the act, rewarding businesses with visibility rankings if they have a mobile site as well as a full sized website.

The Internet & Traditional Websites

The original Internet was a very basic system built in the 1950's known as ARPAnet. It was used for information sharing between universities, the government and relatively a few others, but the various networks were not entirely compatible. Established in 1962, with the first exchange of memos between colleagues, one of which was J.C.R. Licklider of MIT, the network of networks evolved and began to grow. Very few people outside of academia and government were even aware of this burgeoning "World Wide Web." No, Al Gore did not invent it; many heard of the Internet before they even knew who Al Gore was.

Email came into early use in 1972, with little change until more than a decade later. The boom of mass public use began in the early to mid-1980's, with near-constant improvements occurring. The Internet went from DOS to Windows in 1985 and Websites were designed (and still are) to be easily navigated with the use of a mouse on a personal computer (PC). This is a very brief history of which the technologies of today got their impetus. Today there are multiple operating systems, code programs for building sites, drag, and drop. Website themes and an extensive list of providers of which to host a personal or company site. There are ways to obtain a free internet site; thousands of templates from a plethora of sources.

The mobile phone came along, and for a long time it was just that; a phone. I remember my first cell phone came in a bag; sort of like a purse with a phone and battery weighing about 7 or 8 pounds. Technology continued to shrink the cells, improve the connections, but the early years were not all that glamorous. The bag phone graduated to "the brick" and continued to get smaller and smaller, while batteries became smaller and more powerful. Then cameras were added, operating systems for email, and search of the Internet. Things have been progressing more quickly than ever before.

Big Disadvantages

1. The traditional website was and is designed to be viewed on a desktop or laptop computer, where the screen is large, making it easy to read the material, view photos, and search information through vast amounts of data. Companies have built websites with as few as one page to as many as more than a thousand, viewable on a large screen. With the mobile movement already operating at warp speed, these mega sites at full size are far too cumbersome to be viewed effectively on a mobile device. They

can be searched on a mobile device, but seeing with ease is not the case.

2. The traditional website sometimes has flash navigation, which is not compatible with most mobile technologies. Some phones can understand flash but more than 100 million iPhones cannot. That is a huge chunk of the marketplace that is entirely left out if the site has flash navigation.

3. External links in content make it difficult for mobile users because 8 out of 10 sites are not mobile-optimized. This creates either a lack of viewing or just ignoring the links altogether, knowing most will not present properly on their hand-held device. Besides, fat fingers with small type add to the challenge.

4. Page tabs open on a desktop unit is no problem. You can have quite a few pages open concurrently, but that is not the case with mobile technology. Pop-up ads and pop-up landing pages, though attractive for lead capture are an immense frustration for mobile searchers. They distort what it is they are searching, annoying and often drive viewers away. The pop-

over Call To Action forces itself over the page being viewed, causing significant navigation challenges most mobile viewers find unacceptable.

5. Load times for large sites takes precious time and data resources of the device. Some desktop sites I have updated for clients used to take upwards of 20 seconds to load. Can you imagine how long the page load time would be on your smartphone?

6. This will not only frustrate the user but will drive them to your competitors. According to a 2012 study by Google, *"nearly 2 in 3 users are unlikely to return to a mobile site where they had trouble and 40% said they would rather visit a competitor's mobile site instead."*

7. Large pages are difficult to navigate and read on a hand-held device. Large pages with large articles or other content are not conducive to developing positive relationships with customers. Mobile users want critical information, and that's it. They will recall the site at home or work at their desktop computer if they want to research, read or view large files.

8. Large images on large pages make navigation almost impossible because the entire screen of their small hand-held device is covered with only a fraction of the picture. These large images are difficult to load, wasting time and money for the viewer. Not everyone subscribes to an unlimited data plan.

9. Scrolling is a challenge because the size of the screen of the mobile device is much smaller than the traditional web page, and therefore, the device has to scrunch all of the data to fit on the screen you are viewing. Being able to read the information is one thing, the ability to see without scrolling is nearly impossible. Horizontal and vertical scrolling to see the content, having to adjust for maligned text and images will certainly drive a potential customer away from your site, never to embark again.

10. Video marketing is fast becoming the norm for today's marketers, but watching significant amounts of video from a traditional site on a mobile device is costly if the viewer does not have an unlimited data plan.

Even for those with Unlimited Data plans, the streaming often gets stalled and jerky making the experience less than grandiose.

Take Care of Your Customer – Have Them Test their Site

If you are an agency, business owner or entrepreneur, you or your customers may feel that their website is sufficient to handle the whole search for their business online and on mobile. When this objection comes up, it is usually about the business owner knowing if adding a mobile site has any value, or if what they have is just as good, thereby avoiding the expense. A mobile site with a strong marketing strategy is priceless.

Rather than attempt to offer features and benefits to try and sell them on the concept of a mobile website, ask them if they would participate in a simple test of their existing desktop site. An honest answer to the test they perform should settle any questions relating to value and need.

The Test

Ask your customer to contact a friend that is honest and candid, and wouldn't worry about hurting anyone's feelings (brutal candor is what we are after). This person should not be familiar with the website in question but be willing to give honest and candid feedback. They do not need to be face to face; a phone conversation will do just fine. Once connected, have them perform some simple tasks on a smartphone. The objective is to gain honest and candid feedback regarding your customer's existing website from someone they know and trust. Friends usually want to help, so don't let asking for this favor stand in your client's way.

The process they are to perform with their friend using their smartphone is:

Have the business owner ask the friend if she/he would follow the instructions they are about to receive, and after understanding them, hang up and perform the tasks.

1. Have the friend perform a search for the site, and once there, locate the address and phone number and ask them to get directions to the business.
2. Find out what the site features as special.
3. Have them locate the company phone number and give you a call.

On the call, have the business owner go over:
- ✓ how the friend felt the test went
- ✓ how well the friend thinks they did
- ✓ how they found the navigation, viewability

- ✓ were the images clear and within the confines of the screen
- ✓ ask them how long it took to fulfill the three simple tasks
- ✓ Ask them on a scale from 1-10, how frustrated was the experience 1- engaging and fun, 10 – terribly frustrating).
- ✓ Ask them if they had to return to your site frequently, how much time they would spend on it.

If the three simple tasks take more than a few seconds, the chances are that you or your customer is losing business.

Any competitor that knows of the lack of mobile optimization of your client's site is winning the mobile search business. Mobile websites have all the needed information you had the friend find is all on the front page, including a click-to-call button.

Ask yourself how much time you will wait for a site to load before closing the search and looking for a competitor? The chances are that if you wait more than 10 seconds, you are ahead of the curve. The average acceptable wait time for a mobile search is less than 6 seconds before moving on to another provider. It is as simple as that; time is money, and if your site takes too long loading onto a mobile device, you are losing customers.

Mobile is Better than Widescreen

It all boils down to statistics, and more searches are performed on smartphones than any other mobile device. Our friends at Google are very interested in everything mobile, and therefore regularly monitor what his happening in the industry. This 2012 study highlights some of the most influencing statistics that solidify the reasoning behind mobile.

-95% of smartphone users have looked for local information

-88% of these users take action within a day, indicating these are immediate information needs

-77% have contacted a business, with 61% calling and 59% visiting the local business

-61% of local searches on a mobile phone result in a call

Google also puts out useful information about mobile search and mobile marketing. Here's just one example.

http://themobileplaybook.com

Talk about instant gratification; solutions at the touch of a button. Mobile users are searching for restaurants, wellness providers, auto shops, pet stores and services like groomers, doctors, dentists; you name it. Banks offer online bill pay directly from their smartphones, balance notifications, funds transfer and even check depositing via camera are now possible.

You name it, and mobile device searches have them covered. Online directories in local markets are making things easy for businesses to "get found" and directories are the latest in the available marketing weapons at a business' disposal.

Mobile Makes Sense

The most important thing about mobile is its name- **It is mobile!** Remember when there was no such thing as a smartphone? Some of you are not old enough for that thought, but it was not instant gratification by any means. Way back in the days of **P.T. I.** (Prior to Internet) and the **B. M. I.** era (Before Mobile Internet), phone calls occurred when we wanted to either take them or make them. In those dark days, there wasn't even voice mail or answering machines. If your secretary, mom or someone else didn't take a message, you waited until the party called back.

"If it was important, they will call back." That was the overwhelming belief of people back then. If you wanted to know about a company, you looked them up in the directory with those big yellow-paged books and read their ad, called them or visited. **Those days are GONE!**

How did anyone get anything done? Can you imagine trying to depend on a dust-covered book from which to develop a relationship or garner information? What took days now takes minutes, and sometimes happens within seconds.

Image Credit: PSG graphics

Today's consumer is on the go from dawn until way past dusk, and the pace is quickening. Mobile devices are growing in popularity, with smartphones set to reach beyond 7.8 billion soon. There are only 7 billion people on the planet. Know anyone that has more than one mobile device? Sure you do.

Mobile Search is Different

Mobile search is far distinct from a traditional search on a desktop or laptop computer. When someone is searching online using a full sized screen device, it is typically research or curiosity-oriented. You might be researching what new movies are at the theater, where can you find a new widget, catching up on the gossip in Hollywood, the soaps, whatever. Students spend an enormous amount of time researching information to help them learn, and people research companies and goods and services all the time. Mobile users, when searching, are looking for accurate information. They want answers in an instant. What they do not want are slow loading times, a lot of unnecessary content, poor navigation and having to hunt for the information they seek.

Mobile users are busy, on the go people in need of directions, a phone number, a restaurant's menu, a company's service options; the sort of information that might take several minutes to ferret out of a traditional site on a smartphone. With a mobile website, essential information is designed to be available up front and center for easy use.

When was the last time you were in your car and decided to check out the new restaurant about which you heard so much? If they have a mobile website, it will take you no more than about 3-5 seconds to see directions to the restaurant from where you are, what their phone number is with a button to call, and even the ability to view their menu. That is just not possible if it is not a mobile site.

With the world moving at speeds unheard of twenty-five years ago, multi-tasking is the norm. Mobile device users always have them on their person or within easy reach. The devices are on all the time and used at the drop of a hat. These are hard-working individuals in search of instant gratification. If your website is not providing what they need, they move on to the next competitor. They will not settle for less than instant; it is the way we live.

Quality not Quantity

Mobile screens are in the 480 pixels width range, whereas a desktop browser window is more than 800 pixels wide and 1280 tall. That is a huge difference. If the content on a typical site is designed to fit on a desktop screen, jamming it into a much smaller one does not make sense. It is about quality not quantity in mobile site design. If a mobile site has more than a few hundred words, the viewer will not stay long. It is just too much copy. The same thing goes for images; too big and the navigation is shot. If the pictures are more than 480 pixels wide, distortion or lack of view-ability will occur. The 300/480 rule should always apply. Don't offer content with word counts above 300 so the content will fit perfectly on a 480-pixel wide screen.

Mobile website design is particular to everything said. Visitors using mobile search are looking for major information, and are not primarily interested in everything a company has to say, show or sell. The task is simple; get the information fast. Designing a mobile website with a phone, directions, and email buttons on the home page can achieve the viewer's goals in an instant. Mobile users do not have to search all over a scrunched site, navigate back and forth until they can find what they are looking for, or leave the site because the information for which they seek isn't readily available.

Your Responsibility is to the Customer

It does not matter if you are a lawyer, salon owner, a marketing agency, a plumber, dentist, doctor, lawyer, Indian Chief or a business consultant helping small business, you have a responsibility to the customers you serve. You serve the community and therefore owe your potential consumers sufficient information by which they can improve their lives. The hand-held market is growing exponentially, with no end in sight. The only thing I can see replacing portable mobile technology is the Tricorder from Star Trek (there are rumors someone is working on it). Mobile will be with us for a while; it's time for everyone to get on board. The world is already aware of Google Glass (computer eyewear that is already selling in beta testing), so one's imagination only limits technology.

Image Credit: University of Rhode Island

Mobile devices are everything from a music and video source to a shopping aid. Users can compare products on their device right in the store before making a purchase. Mobile search is dramatically affecting the distribution of revenues among competitors, so your business customers need mobile as another weapon against the competition. In a study performed by Google in 2010, **79% of smartphone users utilize their devices to help them with shopping decisions, and** 74% **make a purchase because of the information that they got from their smartphone.**

Mobile In-Store Research

How in-store shoppers are using mobile devices

Shoppers who use mobile more, buy more

Image Credit: Google

Mobile users have their devices everywhere they go or close at hand. They are even browsing while watching television, and that is incredible. A 2012 study by Yahoo found that **"86% of mobile Internet users use their mobile device while watching TV, with 37% of those browsing the Internet for non-related TV material."**

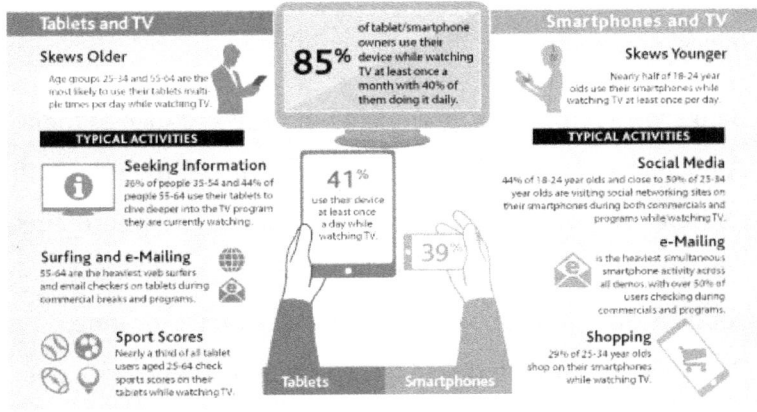

Business Mobile Sites Popularity Up

Mobile Apps Down –

Mobile apps were all the rage a couple of years ago and as is the case with everything, the evolution of technology continues. Don't misunderstand me, mobile apps for business are essential and a tremendous asset for working with their customer base for information, updates, messaging and other sharing.

Banks with apps are experiencing high acceptance of their applications. However, these apps must be platform-specific to their design needs. Banking information and photo-check depositing are great features that require a sophisticated application. Some apps are built for use on the iPhone, Blackberry or Android devices, and they will not work on anything but for which they were designed. We call that "device specific" technology.

Image Credit: Chief Marketer

Developers, therefore, have to build two or three apps for their customers to serve the entire marketplace, as not everyone uses the same style device. iPhone apps will not work on an Android device, and vice-versa. Toss in Blackberry and you have a compounded challenge. Moreover, some Android devices cannot accept some apps no matter the design.

With the ever- increasing use of non-specific web-browsing and search, the "mobile app" doesn't fill the current need at all. Apps must be downloaded before use, so general searches require time and often sign up requirements. Not everyone is comfortable with providing personal information just to get something in return. Apps might serve members of a local group that are frequent users of the business, but those outside the circle of influence will have great difficulty finding them through mobile search.

Image Credit: Quattro

(No Longer Considered the Most Worthy of Investments for the Average Business, Yet Still Important for Some Industries)

Apps on mobile devices were heralded as <u>the</u> things when it came to mobile advertising, but not so much these days.

First, the overall acceptance of mobile advertising has been spotty at best, with consumers disliking banner ads and pop-ups on their searches. Secondly, the actual downloading of apps has declined while the mobile website has surged. Did you know that there are tens of thousands of apps in the app stores that have never been downloaded by anyone? Time, money and dreams wasted on an app that didn't have the appeal needed to be successful.

Mobile App developers are having great difficulty in even getting their app listed in the ios marketplace as well as the Android store. Apple has rejected more apps than ever before this past year, claiming that they **hold little value to the market**. Besides, developers have to pay significant fees just to have their app reviewed. Android does not have the same strict viewpoint, but there are costs involved there as well. Besides, the majority of apps are kept on a device for an average of 88 days. If it is not productivity related for business or personal, the favorite game or gossip site, it becomes redundant within three months. People get bored.

(Mobile Websites-a Must-Have for Today's Competitive Marketplace)

Image Credit: EyesOnDemo.info

Mobile websites can be viewed on any smartphone, bar-none if built on the Eyes On Mobile platform. We guarantee it. Mobile sites are less expensive and quicker to develop than apps, and faster to market is a good thing. Mobile website marketing campaigns can be established in a minimum of time while building a campaign specific app might take weeks or months.

If a small business has neither, the mobile website should be first on the list. Most businesses can do without a mobile app, but the choice is often determined by industry vertical and personal taste.

Once the customer experiences the mobile website, the chances are that an app will no longer be necessary. Apps cost as much as three to five times more than a mobile website and take weeks or months to be perfected. Mobile websites are much quicker to market and cost much less. An uncomplicated mobile site can be built in one day or less and cost as little as a few hundred dollars.

What Google Has to Say About Mobile

Since Google handles more than 94% of all Internet search worldwide, it makes sense that we should listen to what they have to say. They seem to make the rules; control site rankings (at least on their search engine, which is the one that counts) and either praise and reward or punish companies by way of their website search rankings, so we should pay close attention.

Google is all about content, both quality, and delivery. Delivery of engaging and personalized content and not mass produced junk content is number one to Google, and they reward companies by ranking them higher if they have a mobile site as well as wide screen site. The changes to the algorithms Google implemented in 2013 refined search parameters and now award more credit for mobile sites than those found on a PC.

Image Credit: Google http://themobileplaybook.com

It is my belief that Google has it hands down. If more searches are being performed on mobile devices, those in the game should get the credit, while those resisting mobile are being hit with lower rankings overall.

Additionally, Google serves up sites that work the best on the device searched from, so leaner, faster and more user-friendly (mobile optimized) have the advantage. If you find an item of interest on a PC and then a smartphone, you will get different results. That is because Google is paying closer attention to mobile than widescreen; it is the new golden rule. He who has the gold makes the rules, and Google owns the mine.

If your goods and services or those of your customers are not being promoted on a mobile website, you are doing a disservice to yourself and/or your clients. Competition is fierce, and those with the mobile advantage will earn the business.

Check out the competition and see if you are competing on an even keel. It's an easy exercise. Take a look at your competitor's site on your mobile device, then examine yours. Smartphones and tablets have search capabilities, so it isn't difficult at all, as long as you have active Wi-Fi.

Notice if the images are designed to fit well into the handheld screen and don't need to be scrolled to be viewed. If so, the site is mobile optimized; maybe even a mobile site itself. If the mobile site was developed well, special coding has been inserted into the desktop site's computer code. That way when someone searches its URL, the device performing the search will automatically dictate which format to present the viewer. Additionally, a convenient link to the main site should be on the front page of the mobile site to be redirected to the main site. That way the searcher has the best of both worlds.

Remember, a mobile website does not need all of the content of a complete internet site; the Big 4 most important things are location & directions, a phone number, hours of operation and service listings. The rest can be linked to the mobile site to the main site. It is that simple.

Need Versus Want

There are those in this world that seem to think that mobile devices are not necessary and that being able to access the Internet on them is superfluous to living. As far as I am concerned, those people are one of two main types:

1. People that are technology averse (typically called Luddites). These people are usually in the age range of 69 to dead, although there are those that I know in their 80's with a smartphone and some much younger who have thrown off modernity. This group is shrinking due to age-related concerns and peer pressure. I know people in their 40's who are technology averse. They don't have a clue how even to turn on a computer let alone surf the net.

OR

2. People that have yet to experience modern technology due to lack of availability (the sub-Saharan dwellers, those residing in remote parts of India, the Indigenous tribes in the Amazon Rainforest or the Australian Outback, Arctic Nomads). This group is shrinking because of the availability of cheap 3G phones and expansion of the industry to nearly everywhere on the planet. The global market hitting critical mass of this technology will only quicken the pace of acceptance, and therefore the opportunity. The greatest growth of mobile acceptance is in some of these specific locales.

Numbers Do Not Lie

As the number of smartphones in use continues to grow exponentially, those possessing multiple hand-held devices also continue to dominate the world's population to be connected.

Mobile Marketing Association Asia reported in 2011 that there are more people in the world owning smartphones than do toothbrushes. According to the study, of the seven billion people on the planet, 5.1 own a cell phone, 4.2 of them own a toothbrush. Of those 5.1 billion cell phones, more than 3 billion of them are smartphones, with that number growing annually.

Here is a list of statistics taken from **snaphop.com** that you might find as fascinating as do I.

Mobile Marketing Statistics 2013

Mobile Traffic Statistics

Mobile web browsing accounted for 30% of all web traffic in 2012 and is expected to account for 50% by 2014. (Nucleus Research viaSourceCon)

By 2016, the number of mobile devices is projected to surpass the world's population--an 18-fold increase between 2011 and 2016. (Cisco)

Mobile phones accounted for more than 13% *of* all web traffic in August 2012. (Cisco)

Google, Facebook, Yahoo, Amazon, and Wikimedia are the top 5 media properties accessed on mobile devices (comScore 2013)

Mobile Recruitment Statistics

40% of hires come from referrals or job boards (SilkRoad via ERE)

40% of applicants abandon non-mobile-optimized application processes (CareerBuilder)

70% of job seekers look for employment information on mobile. (Simply Hired Job Seeker Report)

55% of today's registered nurses are expected to retire between 2011 and 2020 (BLS via Halogen Software)

60% of job seekers use social media as part of their job search (Simply Hired Job Seeker Report)

72% of job seekers **want** to receive career opportunity information on their mobile devices (Sirona Consulting)

84% of job seekers think job organizations should have mobile-friendly sites (Sirona Consulting)

35% to 40% of job seekers in 10 industries use mobile (CareerBuilder)

86% of job seekers who have a smartphone say they use it to search for jobs (Jibe)

20% of job applications come through mobile for employers with mobile career sites (CareerBuilder 2012)

30% of Indeed.com traffic is mobile from 6 pm to 9 pm (ERE.net)

Mobile Event Statistics

73% of conference attendees make lists of MUST-SEE exhibitors when planning conference attendance (CEIR Study)

Mobile Usage Statistics

27% of emails are opened on mobile devices (Sirona Consulting)

30% of US consumers use mobile devices for shopping (Nielsen Mobile Consumer Report)

80% of physicians use mobile devices at work (Information Week)

Mobile and Small Business

79% of small businesses feel that mobile websites boost customer engagement (eMarketer)

In a recent study conducted by Adobe & eMarketer, they found out that **81% of mobile users prefer mobile sites over apps for price research, 79% of product reviews and 63% for purchasing.** (2011)

Dollars Spent Shift to Online

"Mobile Marketing is the fastest growing industry in history, and is expected to skyrocket to over $50 Billion by 2015"- *IDC Forecast, 2012*

With the passing of time and continued acceptance, mobile commerce is continuing to grow, with revenues annually increasing exponentially year over year. Black Friday online sales in 2015 were nearly half of total sales. This year, experts believe online sales will exceed 50% of the total holiday sales.

The consumer is using mobile for banking, even for the depositing of checks by simply taking a photo of the check within the mobile banking application. Mobile notifications of balance, deposits and withdrawals are commonplace today, when just a few short years ago it was an unheard of concept.

Mobile retail continues to pick up speed, with Black Friday and Cyber Monday events are now happening worldwide, when just 5-6 years ago it was considered "novel." Even mobile credit card acceptance by swipe devices makes digital commerce possible when an electronic register is not available or convenient.

Today's consumer seems always to be pressed for time. Making use of every possible moment employing a hand-held mobile device is par for the course. From banking to retail, appointment setting to reminder texts, all eyes are on mobile. The comfort level of today's consumer is on the steady upswing, with new mobile features and benefits announced regularly.

It is no wonder why mobile is the fastest growing industry in the history of the world. As mobile continues its upward trajectory, this graph shows the increase in mobile buyers annually, as reported by eMarketer.

US Mobile Buyers, by Device, 2011-2017

	2011	2012	2013	2014	2015	2016	2017
Mobile buyers (millions)	34.0	57.0	79.4	98.9	114.9	128.7	138.8
—% of digital buyers	23.7%	38.2%	51.0%	61.1%	68.5%	74.0%	77.1%
Buyers on smartphone (millions)	26.2	41.3	52.3	63.4	73.9	83.0	89.7
—% of smartphone users	29.0%	35.0%	38.5%	41.0%	43.0%	44.5%	45.0%
—% of mobile buyers	77.2%	72.3%	65.8%	64.2%	64.4%	64.5%	64.6%
—% of digital buyers	18.3%	27.6%	33.6%	39.2%	44.1%	47.7%	49.9%
Buyers on tablet (millions)	15.5	50.0	70.6	88.2	102.2	116.5	125.1
—% of tablet users	50.0%	58.0%	63.0%	68.0%	72.0%	77.0%	78.0%
—% of mobile buyers	45.5%	87.6%	88.9%	89.2%	89.0%	90.5%	90.2%
—% of digital buyers	10.8%	33.4%	45.4%	54.5%	60.9%	66.9%	69.6%

Note: ages 14+; mobile device users who have used their mobile device to make at least one purchase via web browser or mobile app during the calendar year
Source: eMarketer, April 2013

154286 www.eMarketer.com

Within the same report, the level of retail sales changing hands from traditional retail to online digital sales is stated by eMarketer, as follows: ***This year's expected m-commerce sales reflect an 11 percent increase from 2012. Total m-commerce sales are projected to reach 25 percent by 2017, topping out at $108 billion in the next four years. This year's $39 billion in retail e-commerce sales is nearly triple the amount earned in 2011.***

For the consumer as well as the retailer, this is great news. Not only is the security of online shopping under constant upgrade, but the accessibility of goods and services has also improved and expanded year over year. The convenience factor alone makes it a sound decision to embrace mobile retail, but also mobile marketing in general. For consumers, the acceptance of digital coupons at the register with a simple reveal of the mobile screen to the cashier saves time and money. As for the retailer, additional opportunities for transactions and customer loyalty occur simultaneously.

As all of the reasons behind shifting to mobile continue in their level of worldwide acceptance, the opportunities for the consumer, the retailer, the service provider and the marketing agency continue to grow.

Additional Marketing Opportunities

As if mobile retail, mobile appointment setting, mobile banking and mobile research and selection are not quite enough to get excited about, consumers are finding that marketers employing some new marketing opportunistic tactics are winning over new customers and clients.

Text Marketing Using SMS –
Short Message Service

If you are not aware of texting, you must be from an alien planet that is well behind Earth regarding technology. What you may not be so educated in is the ability to use SMS messaging for marketing goods and services, improving customer services of businesses and improving customer communications. This happens to be a much more sophisticated program capability than most people recognize. Until someone teaches them about it, which happens to be an excellent way to become that trusted advisor to your customers, she or he will remain ignorant of the benefits of SMS marketing.

Image Credit: Creative Commons

Much like a social media message, the length is limited, so using the right words or a phrase is important, but the functionality is easy to understand and use. You can insert a website link in the message, but remember that your website MUST be mobile-friendly to get a positive reaction. You will frustrate the recipient by offering a link to a site that is not easily visible and readable, so the importance of mobile-friendly is much more critical than simply having a "responsive" site. Responsively designed sites presumably have one believe that it will re-size the site to accommodate the device that has searched for its URL. That is not always the case, however, because there is no hard and fast set of design parameters to govern responsive website design. Not yet.

WORD OF CAUTION: Like email spam, there is also spam in texting. Sending unauthorized text messages is a clear violation of the spam act, so make sure when you establish text campaigns, you have an opt-in system in place.

There is no end to the marketing capabilities of SMS texting, but this book is about why you and or your customers need a mobile site. There is a wealth of information available on our website that teaches and informs about the uses of all of the various marketing strategies, tactics, and tools available so I cannot dwell on this vast scope of specialized training in this book. Check the site for additional marketing information. Better yet, sign up for a free, no credit card necessary trial, our updates and receive relevant information directly in your mail box.

Here's a simple, easy to implement a program for restaurants so that you can begin to grasp the power of SMS Marketing. Like all businesses, repeat business and loyal customers are less expensive to market to than trying to secure a new client. This example highlights the importance of this fact.

Restaurant SMS Marketing Example

Imagine that a pizza restaurant in your town does a lot of print marketing. They offer a variety of coupons in the envelope marketing that arrives every couple of weeks in their customer's mailboxes, the local free ad papers and the like. They also have flyers everywhere in town offering discounts, free deserts, 2 for one deals and so on. Every weekend they have a print ad in the local newspaper, and their monthly budget for these ads is about $1700 a month. Saturday evening is always a slow period for the restaurant, and nothing they have tried has worked to boost business. They experience some business when the youth ball teams come after some games, but that is about it. Imagine you are the owner of the restaurant. Because you learned about the power of mobile, especially SMS texting, you decide to try a new approach.

Because you set up the restaurant with a mobile website and SMS texting, you now have a more affordable way to market to your repeat customers. You can cut back on the amount of print advertising in just a couple of weeks, thereby diverting some of the savings to your new marketing concepts.

Because your restaurant has been offering special "bonuses, running contests" and providing other marketing related messages in all of the advertising are going on, the restaurant has been growing its text database over time, with people opting-in to receive the offers. Now you can begin an SMS Marketing campaign. Sending out promotional texts that people have opted-in to receiving on an infrequent basis (no more than once or twice a week) will grow their business revenues by at least 10-20%. I use the term infrequently because it is not exactly a good deal if they are always receiving a special offer. Besides, too many texts will cause them to opt-out.

Keep Imagining

It is now Saturday morning, and it is raining cats and dogs. The local youth sports teams will not be coming to the restaurant after the baseball games that typically come; the games are rained out. However, your smart idea of sending a text message to all of the coaches offering a Ball Players' pizza deal for the next few hours will fill the restaurant, when it would most certainly have been a ghost town with all the rain. Add to that to "bring the spouse to dinner" specifically for parents and the coaches of the ball players will provide a nice boost in business despite the buy one, get one-half off main courses. You might offer the couples a free, lover's dessert (they share it) of some kind. Sending texts is the least expensive (less than a few cents) method of marketing there is. Think of the savings on your print budget.

A Look at QR Codes

QR Codes or Quick Response Codes have been around for some time, although you may just be noticing them within the past few years. Those people who are aware of them recognize these two-dimensional digital barcodes as a marketing tool. These codes are not winning over fans as quickly as marketers had hoped here in North America, but things are improving with time. What looks like a patch-quilt of black on white can be scanned by a mobile device, and in doing so the viewer can access additional information about a product, service or company quickly. They are good for mobile CTA's (calls to action) couponing, directing them to a particular page of a website, and many other uses.

All smartphones and most other mobile devices with the free downloaded app to scan can capture the digital image of the QR Code. A free program found online can convert the picture to its intended information, or cause a re-direct. There is also software apps that auto-scan and turn the code into a message or link.

Image Credit: Creative Commons

You can use these QR Codes to redirect those that scan it to your website, special offers or any particular product or service marketing of which you can think. As their population grows, you will find them just about everywhere, such as billboard ads, newspaper ads, promotional brochures, even product packaging. Most recently, there was a television ad for a beer company where the young, hip party attendee brings in a 12 pack of brew and scans the QR Code on the packaging. Doing so, he instantly finds out he is a winner of the promotion. The only question I had is why did he wait until he got to the party to scan and not in the store?

QR Codes are great for promotions, events, linking to other online information, unique offerings; just about anything. They can be used for offering discounts such as coupons that can be scanned at the register of any store, and a lot of other purposes.

Here's a QR code for Marketing Partners LLC. Although we are an attractive company with great products and services, our QR code is not very appealing. Things are changing, at least for the QR code industry, and the acceptance thereof.

Here's a sample of the growth of QR code design. Notice the evolution, going from checkerboard to full-color imaging. Cool!

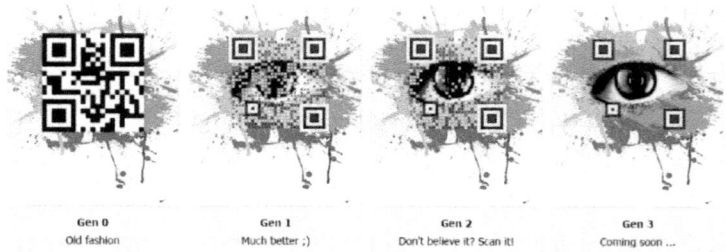

Image Credit: Visualead

With QR Codes having been hugely popular in Japan (they originated this format) for years, they have been a bit slow taking off in the West, but that is changing. As this QR code demo from Visualead and another from Coke show that it is not just a bar code any longer.

Image Credit: Coca-Cola Companies

Here's an example of how to use a QR Code for increasing website visitors.

Have your site URL converted to a QR Code, have it printed on everything you produce on paper. Business cards, brochures, flyers, leaflets, special bulletins, display ads, window signs, posters, even your product bag. Those who scan your code can be immediately sent to your website without having to type anything in the search. That is if they are being forwarded to a mobile-friendly website designed to be easily read and have the information which they will typically be seeking readily available. Business owners that understand the value of QR Codes have a leg up on the competition because they have one more marketing tool in their arsenal than those firms that dismiss the value of this great marketing aid.

If your QR Code sends visitors to a site that cannot be accessed or read on a mobile device conveniently without having to scroll left to right and up and down, you will lose the potential lead and business. The expectations of the segment of the market that utilize technology products beyond a desktop or laptop computer to perform an Internet search will be met if you are part of those businesses that use modern marketing. These folks are looking for instant gratification, which the traditional website on mobile does not offer. Because you have elected to update all of your technology and marketing to meet the demands of the changing marketplace, you stand a much greater chance of sustainability and growth than those on the sidelines.

It would serve no benefit to sending people to a traditional widescreen site that they would have to search endlessly for the information they seek. This would lead to frustration, and eventually create a lasting negative impression of your company and its brand. Mobile search must be maximized for the user because their intentions are much different from those of a widescreen, desktop searcher. Mobile search should yield address and mapping, contact information, hours of operation and special information like services, discounts and the like.

Traditional search is used to find large amounts of information, research, reviews. There is a huge difference in the two types of Internet search users and results expected.

The Marketing of Mobile and the Opportunity

In the history of the world, there has never been a better time to be involved with helping people make their lives better, technologically speaking of course. The truth is that there is a significant shift in the way the entire planet will utilize the digital infrastructure known as the Internet. Mobile search is growing on a daily basis, and we are just getting started.

That means for those who are involved with mobile are destined to succeed, provided the correct approach, service, technology platform and delivery are in place. Opportunity is nothing new; it's the American way. This is the land of opportunity, and it so happens that mobile is exploding worldwide for everyone. This is much more than an American dream come true, it is a global phenomenon building the most powerful technology-related-tsunami in history.

Mobile marketing simply isn't for everyone; opportunity seekers must also be driven entrepreneurs; entrepreneurs with vision and commitment. Marketing Partners LLC has the products, training, marketing materials and ongoing education. You provide the dream, and we will supply the opportunity and platform for free (you must keep reading for more information).

As you can see from the above graphic, mobile marketing is at the center of all marketing. Each marketing segment has a direct connection to mobile. The integration of mobile to search, social media, content, and email is happening and will see a huge jump in popularity in 2014.

My Two Favorite Opportunists

There are two people I highly respect, each alive during 1911 – 1965, but an ocean apart. Not only did the Atlantic separate them, their roles and careers also were so far distant that there is no common link between them except for their outlook on life, and the ability to recognize and understand the opportunity.

 I admire both of these individuals because of their grit, determination, and beliefs.

<u>The Right Honourable</u>
Sir Winston Churchill

Sir Winston Churchill is considered one of the greatest wartime leaders in the history of the world, having served as Prime Minister of Britain from 1940-1945, and again from 1951-1955. **He is the only British Prime Minister in history to have received the** Nobel Prize for Literature **and was also the first person to be made an** Honorary Citizen of the United States. Named the Greatest Briton of all time in a 2002 poll, Churchill is widely regarded as being among the most influential people in British history.

"A pessimist sees the difficulty in every opportunity; an optimist sees the opportunity in every difficulty."

Sir Winston Leonard Spenser Churchill

And…………………..

Lucille Ball – 1911 – 1989

Image and Biography Credit Wikipedia

Ball was dubbed the **"Queen of the Bs"** (referring to her many roles in B-films). In 1951, Ball was instrumental in the creation of the television series *I Love Lucy*. The show co-starred her then-husband, Desi Arnaz, as Ricky Ricardo………   Ball was nominated for an Emmy Award thirteen times and won four times. In 1977, Ball was among the first recipients of the Women in Film Crystal Award. She was the recipient of the Golden Globe Cecil B. DeMille Award in 1979, the Lifetime Achievement Award from the Kennedy Center Honors in 1986, and the Governor's Award from the Academy of Television Arts & Sciences in 1989.

"I do not know anything about luck. I have never banked on it, and I am afraid of people who do. Luck to me is something else; hard work and realizing what opportunity is and what isn't."

Lucille Ball

Two distinctly different people with considerably different occupations, but both with clear vision regarding what it takes to be an opportunist. Both recognized challenges and both sought and seized the opportunity and affected the entire world, each in their way.

Serious Questions for You

Think back 15 years. Go ahead and take a moment to put your thoughts back to the year 1999, and then continue reading.

Ready? The following questions will indicate how technology has changed our day to day lives.

Consider whether you would have believed that people worldwide would be shopping online, let alone from their cell phone? The simple response would be...**No Way!** However, today's answer is a resounding…….**Yes, Way!**

Here's another one for you. Would you have known that this thing called Facebook would dominate marketing by boasting of more than 1.2 billion subscribers? Where people not only migrate but companies of every size and shape have a presence as well? My guess is it would be probably ………**Huh?**

Alternatively still, would you ever believe that the overall population of people under 40 will use texting as a form of communication instead of speaking to someone on the phone? My guess is you would have said…………..**Never Happen!**

What about this…………Would you think it would be possible to see a map, phone number, hours of operation, menu and special information about a new restaurant in town in seconds with the push of a couple of buttons while walking down the street? You would say………. **That is Crazy!**

Would you ever guess that businesses would have the ability to communicate directly to their customers in REAL-TIME on social media because of mobile devices? ………..**No, again!**

One more question to prove my point. Did you ever think that technology is at work replacing the mobile phone and tablet with a pair of visual and hearing device that look like glasses?..........**That Can not Be!**

This is just the tip of the iceberg considering the power of technology **and** mobile. We are at the forefront of the largest shift in consumer –business communications in the history of the world. This is truly an exciting time to be involved with mobile.

Preparedness Meets Opportunity

As in all things in this world, there are no guarantees. For someone to be successful, they need the passion, determination, and willingness to succeed. That is not any great news, but what few people ever consider is the following: To be successful in mobile marketing, they need the right platform from the right partner, the proper training, excellent support, and especially good timing.

Today is the right time for mobile. Mobile search and mobile websites are the foundations of mobile marketing, and they are here to stay. The technology is powerful and getting even better. The Eyes On Mobile CMS platform is the world leader in quality and ease of use. Drag and drop functionality makes DIY so much fun. Of course, we can build one for you if so you choose. Customers in more than 20 countries are proof that the best mobile sites are built on this technology.

Available Here and Now

You may be thinking that mobile websites are difficult to develop and time-consuming projects, but the truth is quite the opposite.

What started out as an adjunct to the traditional widescreen site has been transformed into its industry, and the technology allows websites to be crafted quickly and professionally in little time. Deployment to display correctly on every mobile device in the world is child's play because of our proprietary CMS platform. Whether it is an iPhone, Blackberry, Windows Phone, Android and all matter of tablets, the technology is readily available. Our drag and drop software makes it effortless; no need to understand computer code.

Apps cannot claim this fact regardless of what anyone tells you. Apps do not solve today's business needs when it comes to mobile technology. Apps are often very expensive, and the average length of time a user keeps an app on their mobile device is less than 100 days. Apps work well for games, but not for most businesses other than banks and watching the stock market.

There is no longer need to spend weeks or months building a mobile app and then having the time, trouble and expense of submitting it to the "App Stores," only to be rejected. You will not have to wait for weeks to see if the App Stores even approve your project for distribution. Today's premium mobile websites are customizable, individualized and can be built either by yourself or with us; DIY or DFY. The choice is yours.

Mobile technology continues to expand, and the quality of mobile websites will become ever-more important. With the fact that 8 out of 10 businesses yet to have a mobile site, the opportunity is nearly endless. Now is the time to jump on this opportunity before your competition gains your share of the business.

As business becomes more educated on the value of mobile, the market acceptance and desire to compete will grow along with the need to remain competitive. Those businesses without a mobile website will lose a portion of the market to those companies that understand, utilize and promote their businesses through mobile technologies.

In conclusion, the time is now for getting involved with mobile websites and mobile marketing. **Mobile Search** is the way to Mobile Gold. You and your customers will be happy you decided to become competitive in this terrific technology, regardless of how.

Everyone involved will be ecstatic you chose to work with Eyes On Mobile, a leader in premium mobile websites.

Keep Reading!

Keep reading is some advice my grandfather drummed into my head as a boy, and later as a man. In this particular case, I encourage you to follow my grandfather's advice too. Read on.

At the beginning of this book, I told you, the reader, that if you read the entire tome, you will find a **VERY** special offer. I did not forget to put it in this book, you simply aren't finished reading yet. You are going to like what you find if you are thorough in your education of mobile. Read every page, and you will be rewarded.

About Eyes On Mobile

Eyes On Mobile, one of the Marketing Partners LLC brands, is a worldwide leader in providing the latest premium mobile website software platform ever created. Our Mobile CMS utilizes bleeding-edge technology, and produces the finest looking and functioning mobile websites available anywhere in the world. Regardless of your location, our platform delivers premium mobile sites that drive revenue, improve communication and customer loyalty, and allows anyone to be in business for themselves, but not by themselves.

Eyes On Mobile is a full-service mobile marketing company that can help increase your leads and customers using the latest mobile marketing tools and technologies.

 **Eyes On Mobile** is a team of mobile marketing professionals dedicated to helping business' target their mobile audience to help increase sales and get better Return On Investment (ROI) for your marketing and advertising dollars.

 **Eyes On Mobile's** proprietary solutions can help your business target your mobile-savvy audience, which in turn will contribute to generating more business and increase sales.

Eyes On Mobile offers a full suite of services including mobile ready websites, QR code marketing, and SMS text message marketing as well as small business marketing consulting services. Additionally, if you do not presently have a desktop website, **Eyes On Mobile** can produce a great site for less time and money than all of our competitors, guaranteed.

Eyes On Mobile is one of the several proud brands of the Marketing Partners family of companies. Marketing Partners LLC is a Nevada Limited Liability Company. As a digital marketing agency, our suite of services includes much more than mobile and desktop websites, QR Codes and SMS messaging.

 **Eyes On Mobile** helps businesses, marketing institutions, and individual entrepreneurs make the most of mobile marketing. We offer affiliate programs as well as a white label solution.

 Contact **Eyes On Mobile** today to have one of our mobile marketing experts explain how we can help your business grow using mobile.

Hey Marketers - Seeking an Unparalleled Opportunity?

Read Just a bit further!

If you have read this far, you are serious and should be rewarded. This offer is not mentioned anywhere on the site and is only available to deserving individuals and companies interested in growing a new stream of revenue, or business in itself. By reading this book cover-to-cover, you have demonstrated a sincere desire to learn all you can about the mobile opportunity. I am proud of your dedication to the concept and am happy to provide an unheard of offer. *Keep reading, you are almost there.*

Eyes On Mobile specializes in offering the most robust, full-service mobile marketing platform available on the market today. We help marketing agencies and local marketers set up their own completely white label reseller program, and have resellers in more than 20 countries, selling in more than 17 languages (yes, our entire platform is written in 17 different languages including Chinese. Right to left written languages are in the works, and will be rolled out once they are vetted.

Our mobile CMS provides a cost-effective product that delivers. Our premium mobile sites are easy to create and maintain, so much so that your customers can keep it updated by way of their access code to the back office of their site if you choose them to do so. You will not see any information about our White Label program anywhere but in this book, so contact us with questions about offerings and pricing.

> For the budding entrepreneur, marketing agency or sophisticated local marketer, the proprietary CMS software platform is just the thing to bring premium, custom-crafted mobile websites to your customer base. Our QR codes, SMS messaging and consulting services can set you apart from any so-called competition.

Congratulations!
You made it to the Offer!

Very Special Offer -

Entrepreneurs & Marketing Agencies

Because you have proven you're serious about premium mobile websites (you must be if you're still reading), you qualify. J**ust use the contact us form** on the site **placing one of the codes listed on the next page in the Subject Line to receive a FREE** DIY (Do It Yourself) Premium* Mobile Website, a $697 value (does not include monthly subscription).

If you use the Extra Special Code **in the Subject Line of the contact form letting us know you want to learn how to sell mobile sites and will get access to all our training videos and marketing materials that you can personalize for your own marketing purposes.**

*does not include directory or e-Commerce sites.

As an extra-special offering for both options, receive a 30% discount on hosting, upgrades and maintenance FOR LIFE **for your company site (for a**

Premium Site, normally $97 is just only $ 67 per month for life if you choose to join us.

This offer is subject to revocation at any time.

As I mentioned earlier, **THERE IS NO OBLIGATION. The Trial is Truly Free!**

We will not require you to supply any credit card, just sign up using the contact us form ONLY! DO NOT SIGN UP FOR A PLAN FIRST. DOING SO WILL VOID THE OFFERS!

Usually, we do not provide a free trial, but for anyone using these purchase codes, we are offering a FREE TRIAL of 30 days.

If at any time during the free 30 day trial period you decide you are no longer interested, feel free to close your free account, and you will receive no further communications. If you forget to cancel, no worries, your subscription will automatically cease with no obligation. Remember, you pay nothing until you decide to move forward.

You are not going to find this particular offer on our website, posted in our blog, pay per click ad or anywhere else on the Internet; this limited time deal will disappear without a trace when the mood strikes me. If you are curious, or the least bit interested in learning more, open a free, 30-day trial and see if it is for you or your clients. There is no credit card required until you decide to move ahead.

Special Offer Codes

Special Code
Subject Line – *"Free Site."*
or
Extra Special Code
Subject Line – *"Entrepreneur."*

Remember, USE THE CONTACT US FORM ON THE SITE.

Business Owners - Boost Your Business

Sign up for a free trial and see how easy it is to design a complete, custom mobile website. You will not need any coding experience, any child can drag and drop information, especially if they have the training manual and video collection helping them.

A single mobile site for your business is the best way to move ahead of your competition.

Entrepreneurs - Earn Before You Pay

There is no better way to build and sell mobile sites before you even have to pay anything. As part of the offer, you will have complete access to our exclusive marketing materials that you can personalize, along with our entire catalog of training videos.

Build sample sites and offer them to your prospects. When you successfully collect the deposit or full amount, you are in business! How many opportunities are out there that give you such a big leg up with no up-front cost involved? Not many I am sure. Just contact support, and we will help you get up and running in no time.

You also have the ability to build as many sample sites that you wish to demonstrate for your clients at no cost. They will not become active until you choose to open a paid account.

Cash in on the Gold Rush**.**

Your customers are mobilized**. Are you?**

ABOUT THE AUTHOR

David J Dunworth, Founder & CEO, Marketing Partners LLC

David, **The Over-Caffeinated Entrepreneur,** is a Local Performance Marketing Strategist and Consultant. As CEO of Marketing Partners LLC, a Nevada Limited Liability Company, he specializes in local lead generation strategies. With numerous subsidiaries, he remains quite busy despite being "retired." His areas of expertise are well sprinkled with awards, recognition, and international acclaim. As a serial entrepreneur, he has built and currently manages multiple enterprises, is a published author of hundreds of business articles and three other books. Mr. Dunworth is also a ghostwriter for companies and celebrities alike. A professional business coach and presenter, he has enriched the lives of thousands, offering insights and recommendations to businesses large and small alike. With a global client base, plays a vital role in helping businesses "tell their story" in compelling, engaging content formats, and producing leads for its customers with ease. From mobile website development, digital media to multi-media distribution, local lead generation, publishing and content creation, he, his staff and companies support inbound marketing agencies, their clients and individual enterprises in the UK, Europe, Japan, Africa, India, Australia, Canada and the United States. A believer in focused discipline, tenacious research, performance driven content creation, and unwavering customer loyalty.

Digital Millennium Copyright Act

NOTICE AND PROCEDURE FOR MAKING CLAIMS OF COPYRIGHT INFRINGEMENT

Pursuant to Title 17, United States Code, Section 512(c)(2), all notifications of claimed copyright infringement against Eyes On Mobile, a wholly owned subsidiary of Marketing Partners LLC. ("COMPANY") system or Website should be sent ONLY to our Designated Agent.

NOTE: The Following Information is provided solely for notifying COMPANY that your copyrighted material may have been infringed.

WE CAUTION YOU THAT UNDER FEDERAL LAW, IF YOU KNOWINGLY MISREPRESENT THAT ONLINE MATERIAL IS INFRINGING, YOU MAY BE SUBJECT TO HEAVY CIVIL PENALTIES. THESE INCLUDE MONETARY DAMAGES, COURT COSTS, AND ATTORNEYS FEES INCURRED BY US, BY ANY COPYRIGHT OWNER, OR BY ANY COPYRIGHT OWNER'S LICENSEE THAT IS INJURED AS A RESULT OF OUR RELYING UPON YOUR MISREPRESENTATION. YOU MAY ALSO BE SUBJECT TO CRIMINAL PROSECUTION FOR PERJURY.

DO NOT SEND ANY INQUIRIES UNRELATED TO COPYRIGHT INFRINGEMENT (E.G., REQUESTS FOR TECHNICAL ASSISTANCE OR CUSTOMER SERVICE, REPORTS OF E-MAIL ABUSE, ETC.) TO THE CONTACT LISTED BELOW. YOU WILL NOT RECEIVE A RESPONSE IF SENT TO THAT CONTACT.

Written notification must be submitted to the following Designated Agent:

> Marketing Partners LLC
> 4730 S Fort Apache Rd # 300
> Las Vegas, Nevada 89147
> David J Dunworth Founder and CEO
> E-mail: compliance@marketingpartnersllc.com

IMPORTANT NOTE: IN THE EVENT YOU SEND US A NOTICE OF ANY KIND VIA EMAIL AND DO NOT RECEIVE A RESPONSE FROM US, PLEASE SUBMIT A DUPLICATE COPY VIA PAPER. DUE TO THE VAGARIES OF THE INTERNET, AND EMAIL COMMUNICATION IN PARTICULAR, INCLUDING WITHOUT LIMITATION THE BURDENS OF SPAM AND THE OCCASIONAL, UNINTENDED EFFECTS OF SPAM FILTERS, SENDING AN ALTERNATE FORM OF NOTICE (VIA PAPER), WILL HELP ASSURE THAT YOUR NOTICE WILL BE RECEIVED BY US AND ACTED ON IN A TIMELY MANNER.

Under Title 17, United States Code, Section 512(c)(3)(A), the Notification of Claimed Infringement **must include ALL of the following:**

1.   Physical or electronic signature of a person authorized to act on behalf of the copyright owner.
2.   Identification of the copyrighted work claimed to have been infringed upon or a representative list if multiple works are involved.
3.   Identification of the material that is claimed to be infringing that should be removed or access to disabled and information reasonably sufficient to enable the online service provider to locate the material (usually a URL to the relevant page).
4.   Information reasonably sufficient to allow the online service provider to contact the complaining party (address, phone number, e-mail address).
5.   Statement that the complaining party has "a good faith belief that use of the material in the manner complained of is not authorized by the copyright owner, its agent or the law."
6.   Statement that the information in the notice is accurate, and under penalty of perjury, that the complaining party is authorized to act on behalf of the copyright owner.
7.   Upon receipt of notification of a claimed infringement, COMPANY will respond expeditiously to remove, or disable access to, the material that is claimed to be infringing or to be the subject of infringing activity, regardless of whether the material or activity is ultimately determined to be infringing; if

selective action is not possible, COMPANY will terminate the alleged infringer's Internet access.

COMPANY will also take reasonable steps to promptly notify the alleged infringer in writing of the claim against him or her, and that it has removed or disabled access to the material (see Sections 512(c)(1)(C) and (g) of the DMCA).

Upon receipt of notice from COMPANY that a claim of infringement has been made and/or that the material has been removed or that access to it has been disabled, the Subscriber may provide a Counter Notification.

To be effective, a Counter Notification must meet ALL of the following requirements:

1. It must be a written communication;
2. It must be sent to the Service Provider's Designated Agent;

3. It must include the following
    a. A physical or electronic signature of the Subscriber; Identification of the material that has been removed or to which access has been disabled, and the location at which the material appeared before it was removed or access to it was disabled

b. A statement, under penalty of perjury, that the Subscriber has a good faith belief that the material was removed or disabled as a result of mistake or misidentification of the material to be removed or disabled;

c. The Subscriber's name, address, and telephone number, and a statement that the Subscriber consents to the jurisdiction of Federal District Court for the judicial district in which the Subscriber's address is located, or if the Subscriber's address is outside of the United States, for any judicial district in which the Service Provider may be found, and that the Subscriber will accept service of process from the person who provided notification or an agent of such person.

Upon receipt of a Counter Notification from the Subscriber containing the information as outlined above, COMPANY will:

- Promptly provide the Complaining Party with a copy of the Counter Notification
- Inform the Complaining Party that it will replace the removed material or cease disabling access to it within ten (10) business days following receipt of the Counter Notice;

- Replace the removed material or cease disabling access to the material in not less than ten (10), nor more than fourteen (14), business days following receipt of the Counter Notice, provided Service Provider's Designated Agent has not received notice from the Complaining Party that an action has been filed seeking a court order to restrain Subscriber from engaging in infringing activity relating to the material on Service Provider's network or system.

CAUTION: Pursuant to Title 17, Section 512(f) of the United States Code, any person who knowingly materially misrepresents that material or activity is infringing, or that material or activity was removed or disabled by mistake or misidentification, shall be liable for any damages, including costs and attorneys' fees, incurred by the alleged infringer, by any copyright owner or copyright owner's authorized licensee, or by a service provider, who is injured by such misrepresentation, as the result of the service provider relying upon such misrepresentation in removing or disabling access to the material or activity claimed to be infringing, or in replacing the removed material or ceasing to disable access to it.

Find us at

http://MarketingPartnersllc.com/proximity

Eyes On Mobile is a Marketing Partners LLC Brand

Collaborative. Digital. Guranteed.

http://MarketingPartnersLLC.com

www.ingramcontent.com/pod-product-compliance
Lightning Source LLC
Chambersburg PA
CBHW051719170526
45167CB00002B/722